"Talk to me Pet"

The Simple Guide to Animal Communication

By Sarah Berrisford

Published by Epona Equine Reiki

www.epona-equine-reiki.co.uk

Email: sarah@epona-equine-reiki.co.uk

Introduction

This Animal Communication Guide aims to walk you through the exercises that I teach during the Equine and Animal Healing Courses at Epona Equine Reiki and Animal Healing Centre – see www.epona-equine-reiki.co.uk

I would like to take this opportunity to give a special thanks to the animals who have bought communication into my life. My lovely childhood pony Whisper (now that was a brilliant name considering wasn't it!) and the beautiful Pancho who has helped me with so many aspects of healing and communication.

Also to my Grandfather Lawrence Pavey and Grandmother Janice Brown for being wonderful healers in their own right.

There are a vast amount of questions about Animal Communication: Can I do it? How do I develop it? Surely 'little old me' won't be able to do something so spectacular – actually yes you can! Anyone, yes anyone – that includes YOU can communicate with animals and even better, by working with the exercises during this book you can also develop your skills using different types of communication all at the same time – are you ready to open this new chapter? Let's begin.

How is animal communication useful?

There are many things that animal communication can be useful for, including:

- Finding sore or uncomfortable areas within the physical body
- Finding the cause of behaviour issues
- Improving the relationship and understanding between animal and human
- Checking the correct fit of tack/ equipment used
- Finding out how an animal feels about his lifestyle and environment
- Finding illnesses before they physically present themselves

How do we receive information?

Well, this is the big question really! At this moment in time, I don't see how anyone can give a definite and assured answer backed by research and science, however, we do all have our opinions....

Animals can communicate with each other and so many times I hear a person say "I just can't explain this, it is as if my animal knows what I am thinking" – the short of this statement is, that of course animals know what we are thinking, they are tuned in to what is around them and they 'feel' what we are thinking and what we are putting out. For instance, if one is nervous, we are emitting a tense energy, a tense vibration, the animal picks up on this just as we pick up a feeling from the vibration of a song, some songs we love and they make us feel good, whilst others don't resonate with our bodies, they may make us feel bad or down.

So, on a level we can see how a basic communication with animals can be established, however, it can be harder to understand how we could possibly hear words in our mind which the animal is portraying to us – this can sound a bit bonkers right??!!

I feel that the information we receive from animals is translated by our 'very clever' brains. We have been conditioned to shut off this ability, by peers, friends, family and media as it is something that can be seen as socially unacceptable, fluffy or just plain crazy! But hey, you're reading this book, so you must be one of the three!

The best advice I can give you is to just try out the exercises in this manual, be open minded and just give it a go.

The different types of Animal Communication

Clairsentience

This means feeling what the animal feels, just like with a close friend or family member, you may feel their upset, happiness or to take it further actual physical pain in your body.

The feelings that we receive in our body can be more informative than when the animal actually speaks – rather than hearing the words we feel a definite knowing of what the animal is saying.

Clairvoyance

This is where we see pictures, images or maybe even a video sequence in our minds eye.

Clairaudience

Hearing a voice in your head – no this doesn't mean that you are schizophrenic! The voice can be different each time, male or female sounding or you could hear the words in your own voice, all of these are fine; it is just the way that your brain translates the information that is sent to you. You may hear single words or sentences.

Clairgustance

Some people find this easier than others, they can taste what the animal is tasting; it is possible for them to feel that they have something in their mouth without eating anything. Just think of chocolate, imagine putting a square in your mouth and feeling it melt in your mouth – apart from being a good diet aid, this is what a person feels when practising clairgustance with animals, they can taste the apple which is the horses' favourite food.

Clair scent

This is the ability to smell what the animal is smelling, for instance if there is a strong perfume in the household or a smell that an animal isn't keen on, the communicator can tune in to this and smell the offending odour!

Making ourselves approachable

When we communicate with animals we need to work from the heart, animals can feel our energy, compassion and intention, so the more open and understanding we are, the more approachable we make ourselves.

If you think of the approach in human terms, you will already know yourself, that if someone comes into tour life with a massive ego, telling you how wonderful they are, you probably aren't going to have the best connection with this person – their ego is in the way – this is just the same for animals.

I always suggest to persons whom would like to work with animals in this way, to think of connecting with love. We will talk later in this manual about haling with love, but on a simpler basis we can connect with love. This is very simple, when you meet the animal think of all the things that you love in the world, this could be family, your own animals and also the beautiful animal in front of you. Animals respond to love, it allows them to grow and expand in energy. Love also brings an animal to a state of peace and certainty.

Quietening the mind

The first thing we must do in order to communicate with animals is to stop talking and thinking – how are we supposed to receive messages if we have all the babble from our everyday lives running through our heads?!

We need to open our minds and listen, yes just sit quietly and listen. We need to learn to listen without prejudice, without putting our own views on the situation. Just listening in whole to what is being portrayed. Try this with your friends and family members. Ask them to tell you something that has recently happened to them and just listen, don't comment, just listen – when that voice comes into your head, to suggest something you should say, or to help them find their words, allow that voice to pass, you are just listening, taking in the information.

Once we have established what listening really is, we can begin with the first exercise to quiet our minds. This is probably the most important exercise in this manual – if Carlsberg made exercises....

Quieting the mind instructions: to be carried out for a few minutes each day:

1) Find somewhere to sit in a quiet place, where you will not be disturbed. Turn off any phone, televisions or anything that may distract you.

2) Close your eyes and let your hands rest comfortably in your lap.

3) Take a few long deep breaths.

4) Feel your feet firmly connected to the ground.

5) As you breathe in begin to focus on the deepest point inside your body where your breath reaches. Focus on this point for a couple of minutes and as you do so, you will feel your body become centred.

6) As you exhale feel any tension from your body being released. Each time you breathe out, your body becomes more relaxed.

7) Allow your mind to clear. If you see any images or hear any voices just recognise them, say "hello" to them and then allow them to pass on, as if they are just a cloud floating by or perhaps they are on a conveyor belt. These are not relevant here. They can just be quietly ushered away. You have acknowledged anything that your mind or body feels that you need to know and you have allowed that information to pass on by for now.

8) Allow yourself some time now to feel peace in your mind, to enjoy a clear and quiet mind. If any thoughts come in to your head, just allow them to pass by.

Don't worry if you find this exercise hard at first – most people do! Congratulate yourself even if you manage to have a clear mind for just 10 seconds, this is good and can be built upon. Don't worry if things do come into your head, if you worry about this then you will not only be thinking of that thought but also worrying – that's two things instead of just the one!

I say quite loosely here to congratulate your self, when in actual fact it is very important. Whenever your mind or body does something well, make sure you feel happy, smile and say "well done me" – you can even give yourself a clap if you'd like to! When we feel happiness we make the pathways in the brain stronger. So if you congratulate yourself after quietening your mind for just a few seconds, your very clever brain will say 'that's good, let's make this pathway to a quiet mind easier to get to' – see, now aren't we clever human beings!

With practice you will soon be able to 'turn off' the thoughts in your head as quickly as clicking your fingers. If you would like a challenge, try the same exercise with the radio or television on, eventually you will be able to tune out of the noise around you, so that it doesn't affect you. This state of being open and listening, ready to receive is the same state we need to put ourselves in to hear our animals.

Experiences:

On reading this exercise I became aware that I constantly have conversations with myself! Little voices come into my head and I start drifting off thinking about all sorts of things.

At first I found this exercise extremely hard. I perhaps had one second where I wasn't thinking. Something would come into my mind and I would start thinking about work or replaying conversations I had during the day. At other times I would start playing out future scenarios in my mind. It came to the point where I began to believe, that this mind clearing exercise would never work for me.

I spoke to Sarah about this, with ever the calm attitude, she told me I was doing fine. That this 'one second' I was experiencing could be built upon. She told me to sit down, before carrying out the exercise and congratulate myself on all those 'one seconds' I had managed. Those one seconds added together to make thirteen seconds. I actually had thirteen seconds so far, where my mind had been clear. Still not much, but it was an improvement to focusing on just one second!

After congratulating myself, I carried out the exercise again. I am not sure how long my mind was clear for or what exactly happened during the exercise. It was as if I came back from somewhere else. All I was aware of, after the exercise was a sense of peace and tranquillity; that I have rarely experienced. Also the fact that I wasn't thinking about work!

Mark Smith

I found this exercise incredibly simple to follow. It becomes easier each time I try it.

I now use it all of the time, to create peace in my mind and body.

When the kids are making lots of noise playing, I can retreat to a clear mind, so that I can hear them as a faint background noise, instead of being in the forefront of my mind. I now recommend this exercise to all of my friends. I feel it can make such a difference in our modern day lifestyles.

Diane James

The first connection – Opening the 3rd eye with love

The third eye is located on your brow, between your eyebrows. It is our connection to our psychic abilities. The following exercise will help to open your minds' eye whilst bringing about a strong connection with love. Thus in turn helping you to receive pictures, thoughts and feelings from your animal.

During this exercise you will ask your animal a question, keep the question simple and not too deep. Remember, if you have just met someone you will tend to ask simple, easy to answer questions, not go in depth, asking them why they behave in a certain way or why they do certain things that you don't like. Perhaps start with their favourite food or their favourite thing to do.

On the other hand, instead of asking a question, we can just be open to listening, when an animal senses that a human is finally ready to listen, it is amazing how much they will convey.

The animal you are working with doesn't have to be close by. It doesn't matter whether you are sitting next to each other or on different sides of the planet. We can still communicate with them. Much like you can talk on a mobile phone, there isn't anything physically connecting you to the person on the other side, but instead of using modern technology, like a mobile phone, we use the technology within ourselves. A technology so brilliant that we don't even have to understand how it works, yet it does work and the more we use it and become familiar with it, the easier it becomes to use. We can put our animals on speed dial – no need to even press the numbers. In fact, many people find distant communication easier to start with.

1) Make yourself comfortable in your chair
2) Take a few long deep breaths and go through the quietening your mind exercise
3) Once your body is fully relaxed, bring your attention to your heart. Become aware of the strong but gentle beating.
4) Feel as your heart becomes stronger. Imagine a ray of light coming from your heart and reaching out to touch all living things with pure, endless love.
5) Now, with your eyes still closed, look up for a few seconds. This will bring your attention to your third eye. If anything comes into your mind just acknowledge it and let it pass.
6) Keep relaxing and breathing until you can fill your minds' eye with bright, pure, white light.
7) Now picture the animal you would like to connect with, see his beautiful eyes and feel as your eyes rest on his and you connect.

8) Focus on that connection building strength as you allow your feelings of love for the animal and all animals surface and flow towards him.
9) Now silently in your mind ask a question or be open to listening.
10) Your animal may send you a picture of his answer, you may hear him speak in your mind, or you may receive an overall impression of something in your body.

Don't worry if it doesn't happen straight away. Remember, your animal has probably been trying to communicate with you for years and your mind may have been too busy to hear exactly what he has been saying. Your animal may not be expecting you to suddenly open this channel and because of this is can sometimes be harder to communicate with our own animals than someone else's – well for a start anyway....

Experiences

I asked my friend to read out this exercise to me, so that I could follow it without having to think about what I was doing.

I'm not sure what I was expecting, but I do know that the results exceeded any expectations I could have had!

I felt such a deep connection to my dog, Casper.

On the first connection, I didn't receive any literal information, instead I felt the deepest connection, that previously I could only ever have imagined.

It seems that a wall had been broken down and that I suddenly understood Casper.

I can't put into words what I felt, however, tears were flowing, tears of joy and happiness.

Thank you so much!

Susan Lee, Newcastle

I began writing this email to tell you that I had received nothing from this exercise. I had tried and tried and nothing was happening. I felt that either I was doing something wrong or this animal communication lark wasn't for me!

Then something strange happened. As I sat at the computer writing to you, my cat, Mr Higgins, jumped on to the computer desk. He sat directly in front of the screen looking at me. I tried to move him but he wasn't budging.

I looked into his eyes and said "I can't write this email with you sitting in the way". As he stared back at me, goose bumps appeared all over my arms. I heard a voice in my head "don't give up on me so readily" – I nearly fell of my chair as Mr Higgins placed his paw on my hand, which was resting on the mouse and the message came through again!

I don't know how this happened and certainly can't explain it. You may be able to tell that I am still trying to work this out in my own mind! Perhaps I never will and just need to accept that animal communication is indeed possible and even more exciting, is that I can do it too!!

Stephanie King, Sleaford.

Feeling the question

Remember to keep any questions that you ask positive. Animals are much less likely to answer if you ask them a question with a negative feeling. We need to stay neutral, without judgement on animal or owner.

If you were to ask a horse "why do you kick and bite your owner for no particular reason" you will most likely find that he doesn't really answer. As from the question you have already decided that the horse is to blame and is acting in the wrong manner. Instead we can simply imagine in our mind what the horse is doing, without putting feeling or judgement on it and then be ready to listen. Or we could ask "is there anything your owner can do for you?" This question may well in turn answer what we wanted to know from above, as if the horse were feeling insecure or agitated this could well make him bite his owner.

Experiences

I wanted to know if my mare would like to have a foal. I had tried asking, but felt that I was getting in the way. I would have liked my mare to have a foal, but didn't want to force her, just because it was something I wanted. I also worried that because I was unable to have children, my husband and I had been trying for years, that I may be pushing my wishes on to my horse.

Sarah told me to feel the question. To 'feel' what I was asking and to 'feel' the outcome.

I connected with Lady and asked "would you like a foal?" As I did so I imagined that Lady was pregnant, that she was carrying a special life inside her. I then imagined a foal by her side, running around her. I suddenly felt an overwhelming feeling of proudness coming from Lady.

When I opened my eyes, she was standing directly in front of me. Lady nuzzled my tummy and I felt that she understood my situation.

I closed my eyes and saw a picture of Lady with a bay foal. We decided to put Lady in foal, she took straight away and we were both very excited.

I rode Lady during early pregnancy and after each session she would always nuzzle my belly. I really felt that she understood me and maybe even felt sorry for me.

When Lady was seven months pregnant, I began feeling sick. It didn't pass so I went to the doctors. I was five weeks pregnant!!!

Four months later Lady had a beautiful bay colt, she was a very proud mummy. Each day she would come over and gently nuzzle my bump.

Later that year I gave birth to a healthy little boy. I feel that Lady and I have shared something amazingly special. Both boys are now three years old and they too have a special bond.

I'm sure Lady somehow helped me and I will, forever, remember those days when she used to nuzzle my tummy.

Laura Sharp

I had previously tried to communicate with animals. Having read different books and listened to communicators methods.

Although I felt that I received information, I also felt that something was missing.

A friend of mine shared some advice with me, that Sarah had shared on an Animal Healing Course. She suggested that I should focus on feeling the question and simply listening. Instead of asking lots of different questions.

This really works, I receive so much more by working in this way.

Linda Browne, Shropshire.

Clearing negative thought patterns

Many of us grow up being conditioned to believe in certain things. The problem is, that when we choose to believe in something new, it can open up a whole new world, which in turn can be quite scary – how is it we never knew this was possible? We may also begin to question our beliefs in other areas of our lives that in turn become questionable.

Communicating with animals could be classed as one of these instances and so we may block our abilities for communicating, in fear of what it may lead to for the rest of the beliefs in our lives.

Most people will have some self-doubt in their ability to communicate with animals, this is completely normal.

As much as people would love to communicate with animals they also have an underlying fear of what they may hear. When we are communicating with our own animals, we have a much deeper emotional and perhaps slightly needy connection. When we worry about our animals, we can subconsciously block the messages from getting through to us, safe guarding ourselves from what we may not want to hear.

Experiences

When I first read the instructions, I thought, this will be easy. I am a positive person, I may as well skip this section. I carried on reading to the next exercise Sarah had sent through to me and thought I'd practice that instead. After all, I'm not inundated by negative thoughts.

Maybe it was because I had read the 'Clearing Negative Thought Patterns' and a part of me had decided that I needed to work with this, as later that day I became aware of the niggles that entered my head.

Here are just a couple that came to mind that afternoon:

1) Oh great, I've got to put petrol in the car – more money being spent.

2) Hating this cold weather

3) Why hasn't the husband taken me on holiday this year – doesn't bloody care does he?!

As each of these negative thoughts sprung to mind, I actually began to get the giggles. Only a couple of hours ago, I was thinking just how positive I am and now, here I am, realising just what I am thinking!

I reread the clearing out negative thought patterns exercise and decided to give it a go.

Here's how I worked through my thoughts:

1) Petrol in the car – more money being spent. I asked myself "what really bothers me about this wasting money on petrol" I began to think that I should walk to more places and save money. I could feel a guilt in my left shoulder. Like an aching guilt residing there. I focused on this and said "I'm sorry, I love you".

I began to think 'why do I feel guilty, where does this feeling come from'… ah, it is because I have always been told not to waste. Waste not want not as my father used to always say.

I realised that it is not a waste, if it makes my life easier, I deserve it. I began to smile at the thought of putting petrol in my car. Putting petrol in my car means that I could travel anywhere in the world!

2) Hating this cold weather – it makes everything feel so depressing. I like the heat of the sun.

   I could feel my head become fuzzy, as if I was about to get a headache whilst I thought about this,

   I said to the feeling "I'm sorry, I love you" and then I said to the weather "I love you too!"

   The sun suddenly broke through the clouds in the sky and shon in through the window!!! I laughed out loud, the weather isn't that bad, we still have nice beautiful days in the winter.

3) Why hasn't my hubby taken me on holiday – he doesn't care about me working at home all day, doing all his washing, cooking him dinner, looking after the kids… the list goes on…

   I realise I am feeling unappreciated, I am aware of a sharp pain in my pelvis, again, I say "I'm sorry and I love you" to the feeling.

I still felt a bit sorry for myself, so I chanted the phrase for a few minutes.

I became overwhelmed with compassion for my husband. He is such a hard worker and does so much for the whole family.

Then before I knew it, I was sitting at the computer booking a mini break. I rang my husband and told him. He was thrilled! He had really wanted and needed a holiday, but had been so tired that he couldn't bring himself to start looking into it and sorting out the details

This was a wonderful exercise and we had a lovely holiday!

Kara Blackburn, Northamptonshire.

The ego

Once we have begun to communicate with animals, we also need to check what our ego is doing. For those who find this process very easy straight away, it could be easy to become swept away in the excitement and feel very good about ourselves. Feeling good about ourselves is great, so long as we don't go overboard! Remember, anyone and everyone can communicate with animals, no one is 'more special' because they can communicate quicker or better. We are all special in our own right, and none of us, regardless of our ability is any better than another in the grand scheme of things.

If a person does become egotistical about their abilities, perhaps putting themselves on a pedestal, the animals will see this straight away. Most animals will find this very off putting and an essential line of communication will be lost. Information will still be received but the animal will not open up fully.

We also have to deal with the negative ego. This is the self-destructive part of ourselves. The part of us that whispers in our ear, that we aren't good enough. That what we are trying is for the gifted few and there is no way in which we can achieve it.

If we allow the negative ego to ramble on and on, telling our subconscious mind that we can't do something, then eventually our subconscious will begin to believe it.

We have to remember that the ego is part of our make-up, there is nothing we can do to get rid of it, instead we need to recognise when it surfaces and quieten it down!

To recognise when our ego has come forth, we need to listen to everything we think about ourselves which is negative or upsetting and dismiss it. If you believe you are a failure, you need to delve into this belief. Why do you believe that you are a failure? When was the first time you felt like a failure? What were people saying? What other experiences have you had in your life to reiterate the fact that you're now a failure? Now allow yourself to go through all of these instances - are you sure you still believe?

Every time you hear your ego say something negative about yourself or another person, just stop for a second, recognise that it is your ego talking, like a cheeky child piping up and then laugh at it. You may also like to put something positive in place of the negative thought, for example, if your ego says “you can’t do that”, you may like to change it to “actually I am doing the best I can, and that is fine”.

To start with, the ego may fight back and surface even more regularly, but given time your subconscious will gradually find a different way of thinking, the ego will take a back seat and only come in to play when it is useful to you.

Receiving messages

When we receive information from an animal it is important that we relay that information in the exact same format that we received it, without putting our own thoughts, feelings, guidance or expectations on to the information.

Here is an example of just how easy it is for this to happen:

I was asked to do a communication with a horse that had become flighty when planes flew over from above or if he caught sight of something out of the corner of his eye.

On connecting to the horse, he told me that he had been in an accident whilst driving, just before the accident he had been aware of a plane above him, so this had become a trigger response.

My immediate thought was that the horse had been pulling a cart, as to me that is what driving is, when related to horses.

I gave feedback to the owner and wasn't I glad that I had only given the exact words of the horse, not mentioning being in a cart, just using the word driving.

The horse was being driven in a horse trailer when the accident had happened.

This just shows how easy it can be to put our own words onto something, if I had said to the owner that the horse had told me about an accident that had happened whilst he was pulling a cart, it would have made no sense to her at all!

Behaviour problems

This is a big one! People always ask me "how do I stop my cat from scratching the curtains?" or "how do I stop my horse from banging the stable door?" It is all very well to communicate with the animal and ask them "why do you bang the stable door?" and then the animal says "because I want more food" or "I just wanted to".

So how do we deal with this, it is actually very simple. We need to show the animal how we want him to behave. Take the horse above for instance, you will find that the owner is constantly picturing the horse banging the stable door, every time the owner thinks of the animal she/he thinks of him banging the stable door – now what message do you think this is portraying to the horse? Yes, it is telling the horse to carry on doing it, the owner is telling the horse over and over…. and over again to bang the stable door.

We need to catch our thoughts in this situation. First begin to imagine the horse standing in the stable and keeping all four feet on the ground. Imagine that it is a nice feeling when all four hooves are connected to the ground, he feels calm and happy standing there. Now we need to gradually focus on all of our other thoughts, each time you or the owner begin to think of the horse, you need to think of him standing calmly and happily. If you see images of him banging the stable door, simply spend some time until you can mentally erase that picture and replace it with the desired outcome – now we are sending the correct information to the horse.

If we have been focusing on a horse banging on a stable door for the past three years, it is a bit much to expect him to just stop at the first sign of our 'good' communication skills. Give him time. Improvement is usually seen within a couple of weeks. Sometimes owners just get to the stage where they think it will not make a difference and it is not working, they are ready to give up and then all of a sudden a mini miracle occurs.

Experiences

My two cats, Rolo and Galaxy, had become unhappy with each other. Galaxy always seemed to be moody with Rolo.

I connected to Galaxy to see if I could find out what the issue was. I didn't receive much from Galaxy, just that she found Rolo annoying, he irritated her.

I thought I would see if Rolo would be more helpful… oh boy he was! I felt a weight lift from him as he was able to 'voice' his issues. He told me that Galaxy wanted kittens (Rolo was castrated so that wasn't going to happen!). This explained her behaviour.

Rolo showed me a picture of how he wanted the two of them to be close again.

I decided that I was perhaps going about this in the wrong way. I couldn't do anything about the kittens, but I had to show both cats what I wanted.

Each morning, whilst the cats ate their breakfast, I sat down, closed my eyes and imagined that Rolo and Galaxy were getting on. I visualised that same picture Rolo had sent to me when we connected.

After four days I could see an improvement. Ten days later they were getting along again!

This is all I could have asked for and more!

Tracey Blake, Cheshire

Our family had decided to get a new dog. We already had two dogs, but we had plenty of space and lots of people to share love.

Although I wanted another dog, I couldn't help but worry that this could affect the balance between our two current dogs, who got along really well. We had decided to take on a rescue dog and I know that these can come with their issues.

The evening before our new companion was set to come, I sat down with both current dogs. I spoke to them, explaining the situation and said “I hope you will all get along”. I also didn’t want to let the rescue dog down.

I closed my eyes and imagined all three dogs loving each other. In this moment I felt a sense of calm, like the dogs were saying, ‘all will be fine’.

The next morning Patch arrived. We thought they’d need a bit of time to settle in, how wrong we were. They all became instant best friends. Off they went running in our back field, showing Patch our home.

Later that day they all snuggled on a blanket together – lovely!

Katie Shaw, Boston

Feeling what the animal feels

This is a useful one! We can connect with the animal and feel what is going on in his body, through our own body.

When we are around the animal the pain that we will feel from this exercise is the animals’, it is not our own. When you walk away the pain will be left behind.

In the unlikely event that you still feel pain in your body when you walk away from the animal, just sit down and say "I am ready to release this pain now" as you do so, swipe with your hand over the affected part of your body.

You can also switch this ability on and off. The more you practise this exercise, the more your intuition will take over when you come into contact with animals. This means that when near an animal you will begin to automatically pick up their symptoms. If you would prefer to switch this off you can simply sit down, close your eyes and imagine a light switch in your mind. This light switch relates to feeling an animals' pain, you can turn it off or on whenever you wish.

1) Stand near the animal, somewhere safe, that you feel you can close your eyes.
2) Close your eyes and take a few deep breaths
3) You are now going to work your way through your body tensing and relaxing each muscle to become aware of any pain or tension within
4) Starting with your head, breathe in and tense your facial muscles, hold them there for a second and then when you breathe out let them down and relax.
5) Move on to your shoulders, breathe in and tense your shoulders up, hold them there for a few seconds and then exhale and allow the, to relax down.
6) Breathe in and tense your arms right down to your fists, hold them there and then breathe out and allow them to relax.
7) Be aware of any pains in your own body and allow them to disperse.
8) Breathe in and tense your torso, hold, exhale and allow to relax
9) Breathe in and tense your pelvic area, hold, exhale and allow to relax down

10) Breathe in and tense your legs right down to your feet, curl your toes, hold them there for a couple of seconds and then breathe out and let them relax.

11) Imagine the animal you are connecting to in your minds' eye and say silently to yourself "allow me to feel your pain in my body".

12) Work your way back through your body now, become aware of any new pains, emotion or tension in your body.

13) This pain is coming from the animal.

It is also possible to use the above exercise for distant communication and with humans. If you decide to use this with humans, please remember that you can switch it off – humans have an innate way of leaving their pain with others!

Experiences

When Sarah first explained this exercise to me, I was a participant in one of her animal healing courses. We students gave each other a quick glance to say "oh, we won't be able to do this".

I'll always remember Sarah picking up on this in the first instance and reassuring us that lots of people think that they won't be able to do it, but why not give it a try anyway. Would it really matter if we were unable to do this? No I suppose not. So off we trundled to try out the exercise...

I could still feel the doubt in my body as we were talked through the exercise.

I felt a connection with Stardust, one of Sarah's horses', followed by a strange feeling in the bottom of my back. I quickly opened my eyes and began to study Stardust to see if she had any obvious issues. Unfortunately she had a rug on so I was unable to look properly.

After the exercise had finished, I asked Sarah about the sensation I had felt.

Sarah smiled, walked over to Stardust and took her rug off. At the same time she explained that Stardust was two and a half years old. She was growing and her hind quarters were much higher than her withers. This really took me by surprise, I was actually astounded and felt so thankful to Stardust for sharing.

I still remember what the other students felt too. Lesley said her teeth felt funny with Zorro. Turns out his teeth had been rasped the day before.

Vicky felt an 'opening of her pelvis' and that her belly felt hung down – this was my favourite, as she was connecting with Summer from the other side of the field and with the fence in the way, we couldn't really see her properly. Summer was heavily pregnant!

This was an awesome experience and it still amazes me each and every time I feel what the animal feels.

Sandra Grey, Reading

I proceeded through my body, tensing and relaxing each muscle, whilst also becoming aware of any pains or tension I was holding.

It was surprising just how many of my muscles were holding tension and I found this very relaxing.

After going through my body, I connected to one of my cats, Tiger. We stared at each other for a few moments, eyes fixed, staring right into each other. I spoke out loud "allow me to feel your pains within my body". I then closed my eyes and began to mentally check each part of my body for new pains.

I could feel an irritation on my shoulder, I wanted to scratch at it. I went over to check the corresponding place on Tiger and sure enough she had a small lump there, which looked and felt like an insect bite.

I do enjoy this exercise, but I did start wondering how it works, I emailed Sarah and this is what she wrote "we all have this innate ability, most of us never realise and this could be due to busy minds. We need to listen to what our body is telling us. People pick up symptoms, feelings and energy all of the time, however, they are unaware of this, partly because they don't expect it, but mainly because they are ignoring themselves, they were never taught to listen to their bodies"

Pat Howard, Yorkshire

What if an animal needs healing? How can I help?

You will perhaps notice from my website address that I am a fan of Reiki, a wonderful, self-healing system which in turn allows the practitioner to help heal others.

I cannot teach you the ins and outs of Reiki through this book, but a basic guide is that after an 'empowerment or attunement', which seems to clear pathways in your body, you follow meditations to allow you to be an effective channel for healing. A set of precepts are followed, which read:

"Just for today,

Do not anger,

Do not worry,

Be humble,

Be honest in your dealings with people,

Be compassionate with yourself and others."

These are rather good principals to try to live by – anyway, enough of my sales pitch, let's get on with the book!

I have included below, some exercises that you can use to help heal others around you, this way if you have communicated with an animal and you realise they are in pain, physically or emotionally, you have some tools to help make them feel better.

Please note, I do not recommend just sending healing from your own body, this can leave the healer feeling very drained, instead these exercises all have a common factor of connecting to a 'power source' so that you are simply the channel for healing to come through, receiving healing energies yourself at the same time – have fun, it is most enjoyable!

Healing with Love

So, we have a few different ways to communicate with animals, but what happens then. The animal tells you that he has a bad leg or he is unhappy emotionally, is there anything we can do to help? Yes, of course!

To send healing to an animal is very similar to thinking a good thought. To keep it plain and simple, all of our thoughts are energy. If we send a good thought about someone, that person will receive the good energy from that thought. The great thing is that we also attract what we put out, so if you are thinking good thoughts about people, you are also attracting good thoughts to yourself – great hey!

1) Quiet your mind
2) Begin to focus on the energy around your heart. Feel your heart beating in your body and feel the energy that is being created around it.
3) Whilst focusing on your heart begin to think of all those that you love in this world. As you do so, imagine the energy around your heart becoming bigger and stronger.
4) Every time you exhale, the energy around your heart becomes stronger, bigger, brighter and full of peace.
5) The energy and love being generated from your body fills the room you are in, settling on every object.

6) Now imagine the love from the animal you are connecting to in his heart.

7) Feel as the two hearts connect with love.

8) Allow the love to spread right through the animals' body particularly focusing on the area(s) where he feels pain.

9) Stay in this peaceful loving state for as long as you feel is required.

Love is a very powerful force, it can overcome so many problems. This makes this exercise a very important healing tool.

When carrying out healing we don't impose our will on what we would like to happen. For example, if an animals' time has come and it is time for him to pass over, we don't send healing to keep him alive. Instead we send healing to the situation, to help and soothe all aspects of the situation and to all those involved. To the animal to help his passing. To the owners' to help their grief and to anyone else involved in the situation.

We allow the animal to use the healing that you are channelling for whatever he needs at that time.

Experiences

A beautiful exercise. The deepest connection I have ever felt and extremely powerful!

Tracey Blake, Cheshire.

I love how my dogs react to this exercise. We have two Cocker Spaniels, Tracker and Chester. Chester is quite excitable and Tracker is the sensible chap. When I sit down, Chester likes to come and join me and have a fuss.

I sat down and began the exercise, Tracker was in his basket, whilst Chester was nudging my hand for a stroke.

I ignored both dogs, just concentrating on spreading love throughout the room. I felt Chester connect and as he did so he went to his basket and promptly fell asleep. There was a very strong connection between our hearts, I felt as though both were as one, part of the same being.

My awareness was drawn to Tracker as he suddenly got out of his basket, went over and sniffed the air around Chester's heart. Tracker then went for a scratch and a roll on the tiled kitchen floor (this is out of character, he would usually go into the field to roll and the door was open had he wanted to do this)

A few minutes passed and Tracker made his way over to Chester's basket. Here he climbed on top of him and fell asleep.

I became aware of a connection now being built between Chester, Tracker and I. The energy was making its way around us and it felt lovely.

As I bought things to a close, both dogs woke up and looked at me. We shared a few moments and I thanked them for their time.

Sarah

I have had some really good healing results with this exercise. A dog who was lame behind, suddenly got better straight away. A horse that didn't like having his saddle put on decided that it was no longer a problem and a dog stopped barking incessantly at people.

The healing energy of love wasn't particularly directed anywhere during the sessions. I simply sent love and received love back.

Claire Brown, London.

Guided hands

Letting our hands be guided by our intuition or our higher-selves can be used for many different things. We can look at blockages and the state of the physical and emotional body.

We can either allow our hands to be guided in person, next to the animal or we can work from a distance in our minds' eye.

1) Close your eyes and take some deep breaths

2) Allow your mind to clear and let your body relax

3) Visualise the animal in your minds' eye

4) Imagine your hands are hovering a few inches above the animal

5) Let your hands drift over the body, just follow where your hands are guided.

6) You can ask your hands to be guided to a place of physical discomfort. Or you could ask that your hands are guided to where a problem is originating from.

7) Hold your hands on the area that you are drawn to allow love and good feelings to flow to this area.

Alternative to this you can move your eyesight around the animals' physical body, taking note of any areas you are drawn to. If you find these exercises difficult, just relax and don't try too hard. The harder you try the harder it is to allow your intuition to guide you. The more you relax and don't worry the easier it becomes.

Experiences

My advice to others trying out this exercise is to let go and not try too hard.

I found myself drawn to the loins' area, by a magnetic pull, as if something was gently pulling my hands along. I also like to just stand back and look at the animal as a whole, allow my eyes to become a little unfocused and then somehow I am drawn to a certain place that needs healing.

Debbie Thorn, Shropshire

I found the guided hands hard. Not much was happening, Sarah said to come back to it later and instead use my intuition in a different way.

I was to close my eyes, picture a silhouette of the animal and think to myself that I would see any healing spots as 'flashes of red' on the silhouette in my minds' eye. This I found easy, it really resonated with me and I was even able to pick up the area where Pancho had a precious injury. This appeared as a faded red, as if to say it was old.

Christine Brown, London

Really enjoy this exercise. I find it simple to follow and it always amazes me how my hands move. Sometimes I also hear a voice in my head say 'knee' for instance, which is where I should go.

Sandra Grey, Reading

Body scanning – becoming the animal

Body scanning is another way in which we can find areas of the animal that are in discomfort. It is important to give all of the exercises a try and give yourself a chance to become familiar with them, to see for yourself, which method you get along with best.

1) Sit or stand comfortably
2) Take a few moments to focus on your breathing
3) Feel your feet connected to the ground
4) Ask your animal permission to mentally see inside his body. If he turns away, try another day.
5) Picture an essence or energy gently swirling in your head, this essence from your body, slowly swirls up out of your crown.
6) The energy begins to steadily make its' way over to the animal, where it enters in to the animals' body, through his crown.
7) As the energy enter the animals' body it begins to spread, as it spreads in to his head, you're head becomes the animals head.
8) As the energy spreads down the neck and shoulders, your neck and shoulders become the animals'.
9) As the energy spreads throughout the body, your body becomes the animals' body.

10) Feel as the energy spreads down the animals' legs and how your legs become the animals'. How does it feel to have your feet on the ground like the animal?

11) How does it feel to see through the animals' eyes?

12) If you were to imagine the animal moving, how does this feel? Does it feel comfortable?

13) When you have gathered the information you require, gradually see the essence of energy become smaller within the animal, see it make its' way out of the animals' crown and back into your body.

This exercise can be a harder one to get your head around – it can seem a bit 'out there', but please do try it, it is quite amazing.

The above exercise can also be used to gather information from lost pets. To see whether they are injured. Also to see what they can see in their surroundings can give you a clue as to where they may be.

Experiences

This exercise was very deep and personal. I really felt that a part of my energy had joined with the horse.

The strongest feeling I had, was when I felt all four hooves connected to the ground. It was as if my feet and hands were stuck to the floor, even though my own hands weren't on the floor – very strange!

Kim Barker, Yorkshire

This is my favourite way to communicate with animals. My body feels as though it has somehow morphed into the animals' shape. I enjoy feeling how the animal moves. Running free feels amazing.

I carried out this exercise with a bird and tried to feel what it would be like to fly. It was amazing and I also had my first experience of looking through the animals' eyes. I could see the fields and houses down below – although putting this down on paper makes me sound like a nutcase!

Diana Barnes, Leicestershire

I found this really effective. My dog had decided that when I called him he would run off in the opposite direction.

When I became Jasper, I imagined that he/I was running in the park and that I called him. I felt dismay and sadness, like a child being pulled away from playing on the swings. I realised that Jaspers' problem wasn't really his problem, it was my problem. All Jasper wanted was more time free in the park. I started getting up half an hour earlier in the mornings so that Jasper could have an extra half an hour in the park. I explained this to Jasper, just sitting down next to him, talking to him.

Now when I call Jasper, he comes running. I think he knows that I am doing my best. I also think the exercise helped me, as I thought that I was doing something wrong and I realise that animals aren't keen on this type of energy.

After the communication I had a much clearer picture of what was going on and I also did what I could to rectify the problem.

Claire Brown, London

Developing the different types of Animal Communication

We all tend to find a certain aspect of Animal Communication easier. Some of us find feeling what the animal feels easy whilst others may see many images in their minds' eye

Below are some example questions that you can practice asking, which will help to develop all sides of animal communication. This is because they generate a certain type of response.

Clairsentience (feeling)

- Have you any aches or pains anywhere?
- How does your tack feel? Is it a good fit?
- Do you have any aches or pains whilst being exercised?
- Do your teeth feel comfortable whilst eating?
- Do you like travelling in a vehicle
- Do you like being at shows or visiting friends?

Clairvoyance (seeing)

- What is your favourite exercise?
- Who is your best friend?
- Do you have a favourite place to sleep?

Clairaudience (hearing)

- Is there a message you would like me to pass on to anyone?
- Is there anything that could make you happier?
- Are you happy with your companions?

- Is there anyone you miss?

Clair ambience (taste/smell)

- What is your favourite food?
- Are there any foods you dislike?
- Are there any smells in your home that you dislike?
- Are there any smells in or around your home that you like?

Natural Communication

Many animal owners already have a natural communication with their companions. Without realising what is happening, animal owners will often speak of times when they 'just knew' that something was wrong. This natural communication can be built on, it is a line of expression that has stayed alive, like the emergency number on your mobile phone, which will always work whether or not you have credit.

Below are some examples of what I call 'Natural Communication'

When a horse 'really' tells you!

I'd like to tell you about Cyrano trying to communicate with me. Whilst busy on the farm with my husband Chris, I heard Cyrano whinnying very loudly and persistently. She was in the field with Domino and Webster. I thought I had better go and see what all of the noise was about and as soon as she saw me, she got down and started rolling, got up again and kicked at her belly and kept looking at her side. I immediately though that she had colic. I shouted to my husband that I was going to prepare a stable to bring her in.

I could still hear her whinnying as I was preparing the bed. I went and bought her from the field and she started eating hay. This confused me so I thought I would check the other two. Imagine my surprise when I saw Domino hung dog by the side of the fence, he did not look well at all!

There was nothing wrong with Cyrano, she was telling me that Domino was not well and doing her best to let me know what the problem was. Domino had a stomach ache.

Danuta Lawson

Each morning I bring a group of four horses in from their grass paddock to a bark area.

On this occasion, I was going to introduce something new. Instead of going into the bark area, they were to go through the bark area and into a bare paddock next door.

As I opened the gate, the horses came cantering through. Pancho was first, he stopped in the bark and looked left and right for his feed. Pancho then stared deeply at me, I said "your feed is in the field" Pancho turned and walked out to the field.

Helena Berrisford

It can be quite amazing just how much an animal will try to tell you what is wrong. I have met many animals who will touch a part of their body that is injured with their nose, to point out to their owner where the problem.

What if the situation upsets me?

You are asked to communicate with an animal, the animal tells you that it is soon to be his time, or that he is unhappy with his owners – what can you do?

When a situation affects you and there is nothing you can do about it, stop trying to heal the situation.

Turn inwards to yourself. What is going on in your own body? Become aware of the feelings in your own body that are being created by this situation. You have now become part of the problem. You are in turn feeding the vibration of the situation.

As you become aware of how this is making you feel, make your way through these questions, where is the feeling located? How big is the feeling? Does it have a colour or a shape? Rate the feeling, how bad is it, with zero being fine and ten being terrible.

Now focus on this feeling, send love to this feeling; say to this feeling "I'm sorry, I love you". Chant the phrase to yourself for around half a minute and the rate the feeling again.

This method of healing yourself also heals the problems in others which relate to that issue. Just by focusing on yourself in your private universe.

Conducting a session

Before we begin we must first decide whether we are going to communicate with the animal in person or distantly. Both will basically bring about the same result, however, there are a few factors you may wish to consider.

Firstly, how cold is it outside? It may be hard to concentrate if the session needs to be conducted outside and it is freezing.

Does the animal you are going to communicate with have any undesirable behavioural issues towards humans? If so it can be easier and safer to do a distant session.

How far away is the animal?

Recording information

When communicating with animals, a lot can be said in the time you communicate. There are various ways to record what has been said.

You could have a question and answer sheet and write down the answers as you go.

A tape recorder could be used so that you can repeat what you are seeing and feeling as it comes through to you.

If communicating distantly you could carry out the communication over the telephone or via email. The problem with writing to the owner is that people can read things in different tones and take things in a different concept and so it is usually best to speak to the owners directly, so that you can be clear on what you felt.

You can also provide the owner with a Review Sheet after the communication, with a basis of what went on in the session. On the other hand it is also nice for you to take one of these home to look back on and perhaps build a case study list.

On the following page is an example of the information included on an Animal Communication Review Sheet, which you may wish to copy and use for your own use.

Animal Communication Review

Owners name: Tel:

Address: Email:

Animals' name: Species:

Breed: Colour: Age:

Any known problems/ illnesses:

Reason for wanting communication:

Questions to be asked:

1)

2)

3)

Practitioners Comments

Communication example

(Names have been changed to respect privacy)

Animals Name: Paddy Species: Horse Breed: Welsh x

Problems/illnesses: hay fever symptoms in spring

Reason for Communication: would like to know what he has to say and if there's anything he wants to tell me.

Questions

1) Is there anything he wants to tell me?

Quite strongly, that he likes doing a buck in flying changes. He says you should "stop worrying about nothing". Life's not going to pass you by, follow your dreams.

2) How does he feel about the jumping accident?

Sally was worried about me. One second I was jumping, the next I was falling. He likes jumping for fun still but prefers small jumps that his owner finds fun also.

He says that his owner has a picture that runs through her mind of the accident, he doesn't like that

3) Is he happy

Only if he wants more food! I am getting "don't ask stupid questions"

4) Is there anything I can do for him?

He's showing me a picture of you and him carrying out perfect canter pirouettes, you are smiling and laughing.

5) Anything else?

He likes being your fun pony. Very secure in himself

<u>Areas found on body scan</u>

Body scan showed up lungs and offside buttock

<u>Owner comments</u>

You have described Paddy very well. He is confident and sure of himself. I always describe him as my fun pony.

He does buck when he does flying changes! We are just learning them and it always makes me laugh but my instructor said I must stop him.

I do have that picture from the jumping accident in my head, I wasn't aware of how often I actually think about it until now.

As you can see from the above, we aren't looking for long essays, just for putting down exactly what the horse has portrayed to you. This may only be one or two things, for example, he may just say "I have a headache" and that is it, or you may receive a bit more information. Trust in that the information you receive is what is relevant to the owner at this time.

Being thankful

It is important that we understand that animals don't have to communicate with us, they choose to. This should be reflected in your work with animals by always thanking them for connecting, even if you received very little or nothing at all, give thanks to the animal, this will make them more likely to open up on another occasion.

What happens if the animal doesn't want to talk?

This happens, don't worry about it. If this happens when you go out to see a client, just explain that not much is coming through, thank the animal and tell him that you love him and then go for a cup of tea with the owner, after ten minutes try again – you may want to do this a couple of times with certain animals.

If working distantly, use the same approach as above, however, you can connect to the animal on different days. When I carry out distant communication, I like to have a photo of the animal, when I first receive the photo I will connect with the animal, say "hello" tell him he's beautiful and send love. I'll then get on with something else for a while and come back when I feel the time is right for the communication to begin.

Some would say this is my intuition telling me when the time is right, whilst others might say that I am receiving a calling from the animal – I like to think it is a bit of both. This is the way that works well for me, you don't have to follow in my footsteps, I encourage you to go out and find your preferred way of working – there is no right or wrong way.

Experiences

I had tried and tried to communicate with my dog Freddy to no avail.

Every day, I sat down next to him and tried – nothing.

After doing this every day for three weeks I decided it was time to give up, why had I bothered with this nonsense?!

That night as I lay in bed, Freddy jumped up at the end and proceeded to walk up to my head.

He nudged my face and licked his paw once.

I looked at his paw and saw that he had something stuck in the pads. I carefully pulled out a very small piece of wood cutting, which had somehow become lodged between two pads.

It wasn't until I was telling my friend about Freddy's paw and how I'd given up on animal communication that I realised what had happened.

I had understood when Freddy told me that something was wrong with his paw. We had communicated on some level. As I thought about this, I realised just how many other times I had 'known' what Freddy had been telling me. We didn't need to communicate in a conversational manner, we had our own language and our own way of knowing what the other species meant.

Diane Holmes, Leicestershire

The owners' influence

Animals see straight through our exterior and into our soul. It is important that we humans learn to open up and be completely honest with our animal companions.

Our animal friends know if we are trying to hide something, whether we are trying to bury emotions or just have busy minds, animals see the energy that we are portraying and they know whether the words coming out of our mouths are matching the energy we are portraying. For example, say if someone's dog is barking at a stranger at the door. The owner may say "shhh be quiet" verbally, however, internally the owner could well be saying "protect me from this stranger" and so the owner is giving conflicting messages to the dog. The dog will more than likely listen to the owners' inner wishes or protection or he may become confused or upset as he feels he is misinterpreting the information being given to him on how to act.

Bearing the above in mind it is very important to listen carefully to animals and their owners when we are trying to communicate and help. As what could be seen as a behavioural problem could just as easily be coming from conflicting instructions from the animals' human.

Here is a letter a client sent me after I had visited her horse. She has given me full permission to share this letter through this book:

Dear Sarah,

Thank you for today, I know Thunder enjoyed the time he spent with you.

I know for the first five minutes he was trying to lunge at you over the stable door, but if you knew what he was like on an everyday basis, you too would be very surprised.

Thunder usually lunges at people constantly; there is no stopping him, so when he suddenly accepted you I was amazed.

I felt so calm and relaxed when you left, I sat in a chair at the yard and began to think about my life.

I realised that Thunder acts in the way that I feel. I just want to tell people to go away and to leave me alone; just what Thunder does all of the time. Unfortunately, I didn't have it in me to do this. Or so I thought…

On Monday morning I went into work. I was sitting at the computer typing away when one of the girls came by. She slapped a piece of paper on my desk and said "this takes priority now". This happens to me all of the time, people off load their work on to me and I let them, just to save any confrontations… well usually…

Something rose up in my stomach, a fire that rose further into my throat. Before I knew it, I had got up out of my chair, picked up the work which wasn't assigned to me, walked over to said ladies desk, dropped it on the side and said "do your own bloody work!"

As I turned back round, I felt the best I had in years! I noticed other workers smiling at me, like I could hear them saying "you go girl".

It now seems crazy that I have let this go on for years, any time someone tries to push their work on to me I just say "I don't think so love!"

As I have found my voice, it seems that Thunder is settling down. He has the odd person that he's not keen on. Most if the time though, he is pleasant and has actually started whinnying at me when I arrive at the yard. This stirs my heart every time.

Thank you so much for your help.

Katie Miller, Cambridgeshire.

A dog is still a dog

A common misconception about Animal Communication is when people think that once they have learnt to communicate, their animal will do as they ask... remember animals still have a choice, just because we have asked them to do something in a way that they will understand, doesn't mean that they will! We can have the best relationship with a dog say, we love him, he loves us and we have a great understanding of each other. Still when said dog catches a scent he is more than likely to follow his natural instinct and see where the scent leads. If we expect our animals to do everything that we say and ask, then I always suggest getting a robot or maybe a teddy! Respect and love from our animals shouldn't be taken for granted.

Extra meditations to help you connect and become one

We all have the ability to love. Showing selfless love is the key to our natural abilities. Instead, however, many of us shut ourselves off and keep ourselves shut down in order to protect ourselves from being hurt. We need to take time each day to sit and listen. Listen to ourselves, listen to our animals and listen to the world around us.

Connecting with the elements

Take some time each day to listen to the elements of nature, to connect with the world around you to tune into Mother Nature.

1) Sit and listen to the wind
2) How does it sound? How does your body feel as you listen? Where in your body do you feel the breeze?
3) Watch how the breeze blows its' surroundings, the trees swaying in its' path. How it travels across the Earth. How it moves the clouds through the sky. Listen, watch and feel.
4) Imagine what it would be like to be the wind. Imagine how it would feel to run freely over the Earth and in the sky.
5) When you have experienced this for a time that suits you, move on to the next step.
6) Feel the heat of the sun on your face. See the light reaching parts of the Earth. How does your body feel as you think of the sun?

7) Imagine what it would feel like to be the sun. How does it feel to be strong and powerful? What does it feel like to pour light on the shadows, to nourish with your warmth?

8) Feel as your rays touch every living thing, as they inspire life and growth.

9) Now take your attention back to the Earth, look deeply into the ground. Rest your hands on the soil or rock – what do you feel in your body?

10) Look deeply in to the core of the Earth, become one with the heat of the core and gradually make your way to the surface. Feel the energy around your surface.

11) Feel as the heat of the sun makes you feel new. Feel the plants and animals on your surface that you have grown and nourished.

12) Now take your attention to the sea. Remember how the waves crash powerfully on the rocks. And then recall how calm the sea can be lapping gently on to the beach. How does this feel in your body?

13) Connect with the sea. Imagine what it would be like to be the sea. Feel the enormity of your size and presence. Feel the nourishment of water.

14) When you feel ready, gradually bring yourself back out of this meditation.

This meditation is lovely when you get the hang of it. Don't give up if you find it hard at first. It is a beautiful experience to become one and gain a deeper understanding on the elements of nature.

Experiences

I love doing this exercise; it makes me feels so much closer to nature!

James Ryan, Nottinghamshire

The wind seemed to be going on forever. I was sick and tired of going outside in these horrible conditions. Sarah said to try this exercise and connect to the wind. It was quite eye opening, somehow it made me respect the wind, respect its strength and also how still it could be.

When I ventured outside the following morning it was blowing a gale. Instead of feeling dismay, I felt empowered. The wind was beautiful and powerful.

Thank you, I feel that this has opened my eyes to another way of working.

Mary Smith, Cambridgeshire

Connecting to your Animal Spirit Guide

Animal guides will come to us when we need them. Sometimes your guide is a pet who has passed over whilst others it can be something totally different. Each animal has their own meaning, however, we don't need to worry about this.

Instead we can use the following exercise to connect with our spirit guide who will help us when we are unsure in a situation.

1) Sit comfortably
2) Take a few long deep breaths and relax
3) Quiet your mind and allow your body to become centred
4) Imagine you are walking through a forest. The forest has a safe, peaceful feeling.
5) In front of you is a huge tree
6) As you approach the tree, you can see a door in the trunk
7) You walk up and go through the door
8) Inside the tree you see a staircase leading down, you slowly make your way down the stair case
9) You can see a beautiful light at the bottom of the staircase, as you walk towards the light is becomes brighter.
10) As you reach the bottom of the stairs you are immersed in a beautiful light. You step out into the light.
11) It is beautiful here, you have found a safe, peaceful and quiet place.
12) There is a lake in front of you. You walk over to the lake where you find a comfortable place to sit down and relax.

13) Sit here and allow an animal to come forth

14) The animal may give you something. Ask the animal if there is anything you can do to help you on your path with animal communication. Ask the animal is there is anything you can do for him.

15) When you have finished, you may like to enjoy the peace of your safe place for a while, or you may like to go back up the staircase and into the forest again.

16) Take some deep breaths and slowly bring yourself back

Don't be surprised is your animal seems like a small insignificant animal to you or isn't what you had expected. This is quite normal.

Legalities

It is important to state to owners that animal communication and healing is not a substitute for Veterinary care and treatment. In the first instance a vet should be called.

We are not allowed to diagnose illness in other persons' animals.

We are not allowed to prescribe treatment to other persons' animals.

Where an animal is ill, Veterinary permission should be sought before hands on healing is used.

Insurance

For those persons who wish to work on the publics' animals. It is important to take out insurance. I would suggest www.therapistinsurance.co.uk

A final thought…

If your animal friend thinks you're the best,

Don't seek a second opinion…

Even from yourself!

About the Author

Sarah Berrisford was born in Woodford, London. Currently living in South Lincolnshire running Epona Equine Reiki, Animal Healing courses, Animal Communication Courses, and Intuitive Horsemanship demonstrations. Sarah also works in the small family stud 'Yeguada Mistral', with their Spanish Stallions.

Sarah believes that all persons can communicate with animals. It is a natural ability and she has taken this further by putting together this simple guide to Animal Communication. We hope you have enjoyed this handbook.

Other publications by Sarah Berrisford

The Handbook of Equine Reiki

The Complete Guide to Animal Reiki

Reiki in the Saddle

Lightning Source UK Ltd.
Milton Keynes UK
UKHW04f1115230518
323079UK00004B/22/P